What Do You Think?

Is There Other Life In The Universe?

Kate Shuster

Heinemann
LIBRARY

www.heinemann.co.uk/library
Visit our website to find out more information about Heinemann Library books.

To order:
☎ Phone 44 (0) 1865 888066
📠 Send a fax to 44 (0) 1865 314091
💻 Visit the Heinemann bookshop at www.heinemann.co.uk/library to browse our catalogue and order online.

Heinemann Library is an imprint of **Pearson Education Limited**, a company incorporated in England and Wales having its registered office at Edinburgh Gate, Harlow, Essex, CM20 2JE – Registered company number: 00872828.

Heinemann Library is a registered trademark of Pearson Education Limited.
Text © Pearson Education Limited 2009
First published in hardback in 2009

The moral right of the proprietor has been asserted.

Editorial: Andrew Farrow and Rebecca Vickers
Design: Philippa Jenkins
Picture Research: Melissa Allison, Ruth Blair and Virginia Stroud-Lewis
Production: Alison Parsons

Originated by Heinemann Library
Printed and bound in China

ISBN 978 0431 11192 6 (hardback)
13 12 11 10 09
10 9 8 7 6 5 4 3 2 1

British Library Cataloguing in Publication Data
Shuster, Kate, 1974-
Is there other life in the Universe? - (What do you think?)
1. Life on other planets - Juvenile literature
I. Title
576.8'39

Acknowledgements
The publishers would like to thank the following for permission to reproduce photographs:
© AKG Images/Erich Lessing p. 7; © Alamy pp. / Chris Cheadle 20, /Mary Evans Picture Library 49, /Jacky Parker 27, /Friedrich Saurer 35; © Corbis pp. 19, /Bettmann 41, /Charles Gupton 50, /Gabe Palmer 4; © Getty Images pp. /Science Faction 23, /Time & Life Pictures 22; © Istockphoto pp. 30, 31; © The Kobal Collection pp. 28, 38, 40, /Allied Artists 43, /Warner Bros 37; © NASA pp. /Q 42, /The Hubble Heritage Team, STScl, AURA 44, /JSC 8; © Nature Picture Library/David Shale p. 36; © Science Photo Library pp. /Christian Darkin 26, /W K Fletcher 33, /Mark Garlick 14, /James King Holmes 32, /Los Alamos National Laboratory 17, /NASA/JPL-Solar System Visualization Team 15, /Gregory Ochocki 34, /Dr. Seth Shoshtak 18, /Detlev Van Ravenswaay 16; © SETI@home, University of California 2008 p. 25; © Kate Shuster p. 44; © Superstock/age fotostock p. 12.

Cover photograph of an artist's impression of an alien spaceship flying above a space shuttle reproduced with permission of © Alamy/Jupiter Images/Brand X.

The publishers would like to thank Dr Geza Gyuk for his assistance in the preparation of this book.

Every effort has been made to contact copyright holders of any material reproduced in this book. Any omissions will be rectified in subsequent printings if notice is given to the publisher.

Table Of Contents

Some words are printed in bold, **like this**. You can find out what they mean in the Glossary on pages 54–55.

> **> Infinite space**

There are trillions of stars in trillions of galaxies throughout
the universe. Many of these stars have planets orbiting
them, some much older than Earth and others forming all
the time. Could there be some form of life on these planets
scattered throughout the universe?

Are We Alone?

If you look out at the night sky sometimes and wonder, "Are we alone in the universe?" you're asking a question that humans have asked for all of history. The universe is so large that it is impossible to imagine its size and scope. Surely among all of those other galaxies, stars, and planets, there are other creatures wondering, like us, if they are alone in the universe? Most people believe there is other life in the universe, even though there is little or no concrete evidence for its existence. We do not even know if there is other life in our solar system, much less whether there is life on distant planets in other solar systems or other galaxies. And we may never know about that other life, because it could take thousands of years to reach by space travel, using technology that so far we do not have.

But humans continue to wonder about life beyond Earth. Images of aliens, both friendly and hostile, are all around us in books, on television, in films, and in video games. The human fascination with **extraterrestrial** life is many thousands of years old, and shows no signs of decreasing. This book asks you to think critically about the question: "Is there other life in the universe?" You will be presented with evidence and arguments that relate to both sides of this question. Your opinion will count. What do you think?

Looking towards the stars

Since the beginnings of human history, the mysterious workings of the cosmos have been essential elements of religion, philosophy, and everyday life. The arrangement of stars was thought to influence everything from crop growth to military success. Most ancient religions had explanations for the significance of the stars and constellations, and it was not unusual for citizens of the great ancient empires, such as those of Egypt, Sumeria, India, Babylon, Assyria, and China, to believe that there was life beyond Earth. As part of religious belief, the usual explanation was that the universe was home to gods and demons who might or might not interact with people living on Earth.

In the sixth and seventh centuries BCE, Greek philosophers, including Thales and Anaximander, argued that the universe was infinite and therefore had an infinite number of worlds with life. This belief was not shared by other ancient philosophers, who subscribed to the **geocentric** (earth-centred) universe theory, believing all life in the universe revolved around Earth. Until people began to accept Copernicus' discovery that the solar system was **heliocentric** (Sun-centred), most people believed that Earth was the unique focus of the universe.

Knowing the universe

As telescopes became more widely available and the visual universe was mapped and considered, more scientists became convinced that there might be other solar systems besides our own, and other planets in those solar systems. Many thought there was likely to be life on other planets in our own solar system. Theories about the nature of life on other planets abounded. Even though there is no evidence of alien life that the majority of scientists accept, many people believe that extraterrestrial life exists. Many people have claimed to see aliens or the evidence of alien visits. Others are convinced that the statistical probability is that we are not alone. Still others are not convinced. They believe Earth is a rare and unique example of a planet that supports life.

✔ The development of the optical telescope

Although the properties of lenses to magnify things had been understood for many years, the first practical telescope was invented in 1608 by the Dutch lensmaker, Hans Lippershey. The astronomer Galileo further refined the instrument and was the first to see and identify the craters on the Moon and sunspots. His first telescope had a magnification of approximately 30 times what is visible with the naked eye.

The trial of Galileo

In 1633 the Italian astronomer Galileo (1564–1641) was put on trial in a church court for his belief in the Copernican theory of the heliocentric universe. The Catholic Church would not accept any challenges to its teaching that Earth was a fixed object around which all the other heavenly bodies revolved. Galileo was forced to renounce his heliocentric opinions and retire from academic work. In 2000, Pope John Paul II formally apologized on behalf of the Catholic Church for the trial and conviction of Galileo.

> *The Sumerian Sun-centred universe*

Long before the early astronomer Copernicus (1473–1543) argued that Earth revolves around the Sun rather than the other way round, ancient civilizations such as the Sumerians believed that the Sun was at the centre of the cosmos. Seals like this one are evidence that ancient civilizations thought about and made images of what they believed the universe was like beyond our world.

What do you think?

This book asks you to develop an informed opinion on one of science's most enduring, controversial, and captivating questions: is there other life in the universe? You may already have an opinion about this subject, or it may be an issue you know nothing about. Either way, developing an informed opinion requires reading different perspectives on an issue. As you're exposed to those different, often conflicting, ideas, you'll be asked to think critically about them.

Critical thinkers set aside their own opinions and ask questions of the ideas and opinions they encounter. This allows them to approach new ideas with an open mind. They might ask questions such as the following:

• Is this information *biased*? Does the author have a reason to state his or her opinion in a particular way, or to show only some of the information?

• Is the information *sufficient*? What other information might you need to make a decision about the issue? Where could you find that information?

• Is the information *credible*? Is the point of view supported by reasoning and evidence?

• Do the arguments rely on *sound logic*? Are there any failures of reasoning that might weaken the argument?

The point of critical thinking and critical reading is to approach ideas sceptically. Paying careful attention to new and different ideas will help you make your own informed opinion.

Forming your own opinions

At first, it might seem that the existence of extraterrestrial life is a matter of personal belief. Beliefs do not have to be supported by facts, and they cannot be proven or disproven. You might have beliefs about extraterrestrial life, but it would be better to have opinions. Opinions can be right or wrong, and they can be proven or disproven, depending on reasoning and evidence. Critical thinkers form opinions based on the available evidence, and they express their opinions by making arguments. An argument has three parts:

• An *assertion* is a statement about the world. It summarizes the main point of your opinion or idea.

• *Reasoning* is the "because" part of your argument. Reasons support your argument or idea with sound logic.

• To back up your reasoning, you'll need *evidence*. Evidence is a fact or example that supports your reasoning, proving that your argument is a good one.

One way to remember the parts of an argument is by the abbreviation A—R—E for assertion, reasoning, and evidence.

As you are presented with different ideas about the existence of extraterrestrial life, you should evaluate the reasoning and evidence you are given. Because there is no generally accepted evidence of extraterrestrial life now available to science, much of the evidence is **inferential**. Inferential reasoning is reasoning we use when we are trying to reach conclusions that go beyond the reach of the available information. For example, if you surveyed ten of your classmates to find out if they like hamburgers, you might be able to infer from the results more general findings about whether your whole class likes hamburgers.

This is called taking a sample from a population. It is one kind of inferential reasoning. Another kind is when we infer ideas about something we do not know about (like alien life in the universe) based on something we do know about (like humans). Scientists who think about extraterrestrial life do a lot of inferential reasoning. In this book, you will be asked to do this, too.

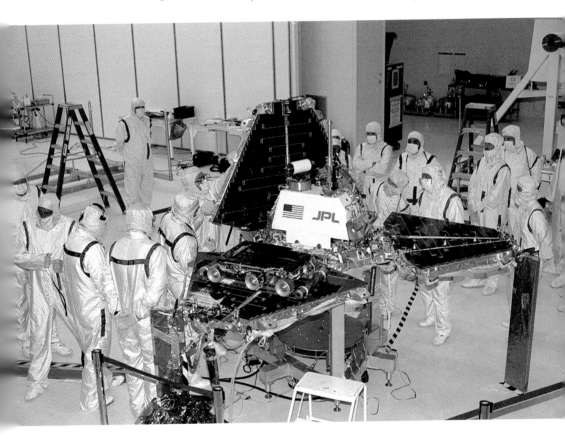

> *Science is a critical thinking profession*

Because scientists try to test explanations, or **hypotheses**, about the natural world, they must collect evidence and examine it carefully. They weigh the available evidence and explanations to decide whether their hypotheses are more likely to be true than false.

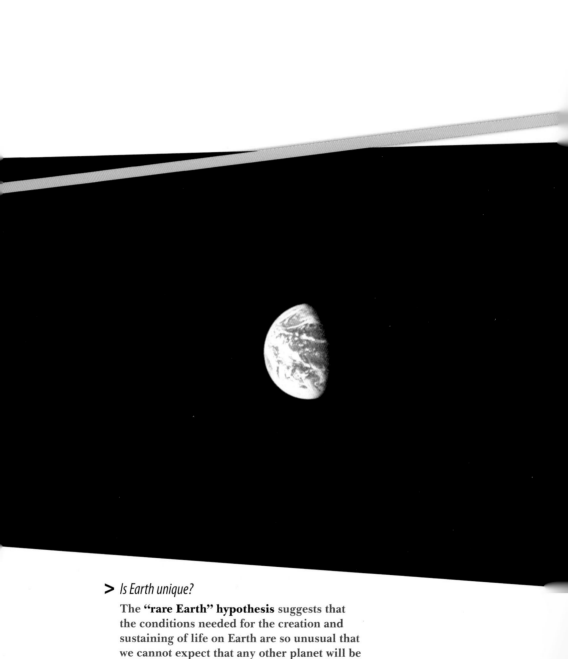

> *Is Earth unique?*

The **"rare Earth" hypothesis** suggests that the conditions needed for the creation and sustaining of life on Earth are so unusual that we cannot expect that any other planet will be able to sustain living things, let alone any as complex and delicate as human life.

What Are The Odds?

As a young astronomer, Frank Drake studied distant galaxies for clues that might help him understand how new stars were born. Late one night, he noticed a very unusual signal from his **radio telescope**. It was like nothing he had ever seen before. He tried to find the signal the next day, but it had vanished. Drake reasoned that such an unusual signal might be the product of intelligent life – that it might, in fact, be the method an alien life form would use to communicate with other life in the universe. Excited by this prospect, he worked with a group of scientists to try to work out how likely it is that other life exists in our galaxy (the Milky Way).

The formula he came up with almost 50 years ago has become known as the **Drake Equation**. It identifies seven key factors that, taken together, predict how many civilizations might be trying to contact us. Dr. Drake has spent his life looking for extraterrestrial life based on his belief that such life is likely to exist. His convictions, and the opinions of other scientists, have been greatly influenced by the factors in the Drake Equation. In this chapter we will survey these factors, ranging from the possibility that other planets exist beyond our solar system to the likelihood that these distant other worlds support life.

How many stars can support life?

The first factor in Drake's equation is the number of stars in the galaxy. Astronomers have estimated that there may be as many as 300 billion stars in our galaxy. Although no one is sure, there may be 100 billion other galaxies. This means that the total number of stars in the universe could be 30 trillion or more. But not all these stars can support life on the planets associated with them. Some are simply too hot, or too cold, or too unstable. Others might be too close to a **black hole**. In order to work out the likelihood of the existence of other life in our galaxy, we need to know how many of these stars could potentially support life.

Stars do not stay the same forever. Like plants and animals, they are born, age, then die. When stars are formed, they are extraordinarily hot. Over time, their temperature changes. Sometimes they may collapse. Even during the main part of their cycle, all stars are not the same. Astronomers remember the seven kinds of stars using the line, "**O**h, **B**e **A** **F**ine **G**irl [or Guy], **K**iss **M**e!"

O and B stars are blue stars that burn even hotter than our Sun. A and F stars are the next hottest stars, and are white. Because they burn so hot, these first four kinds have relatively short life spans. They may last for only a few billion years. The next hottest kind of star is a yellow G star. Our Sun, like five percent of all the stars in the galaxy, is this rare kind of star. After G stars come even cooler stars: these are the orange K stars and red M stars. Both of these kinds of star have less heat than our Sun. If G stars are five percent of the stars in our galaxy, and there are 300 billion stars in the galaxy, then 15 billion of those stars are G stars, and therefore potentially able to support life.

How many of those stars have planets?

Just as not all stars are the same temperature as our Sun, not all star systems have planets like ours. Stars form when dust and gases come together in nuclear reactions that burn and produce light and heat. After stars form, they continue to be surrounded by spinning clouds of matter that was not absorbed into the star itself. This whirling matter is what may condense to make planets. Because planets are hard to see from a great distance, we are not sure of the frequency of planets related to other stars.

It seems likely that nearly half of all stars might have planets of some kind, given the way that stars and planets form. With all that spinning dust, there is likely to be plenty left over for making planets. But nobody is sure what these other planets are like. Are they big, small, hot, cold? It depends on their position in relation to their star, and where their star is in the galaxy.

> *No water, no life*

Most scientists believe that liquid water is essential for the formation and maintenance of life. Humans are 60–70 percent water, and some life forms, such as jellyfish, may be up to 90 percent water. Water is a **solvent** that dissolves liquids and solids, allowing them to be transported into and out of life forms. It also makes all essential life processes possible by aiding **metabolism**, allowing life forms to create larger molecules and break molecules down into smaller parts. Without water, no life as we know it could exist.

Can other planets in the universe support life?

For planets to be hospitable to life, most scientists agree they must be able to support liquid water, which is essential to the development and maintenance of life as we know it. There may well be other forms of life that do not depend on water for their existence. For example, sulphur could be the solvent that enables an entirely different kind of life to form and flourish on some far-away world. But we are just not sure if a sulphur- or silicon-based life form is even possible, so water is still considered essential to the maintenance of life.

Earth may not seem so special to those of us living on it, going about our daily business. But to scientists it is extraordinarily special. It is not too far from the Sun, which would make it too cold to sustain life. It is also not too close, which would make it much too hot for life as we know it.

Earth revolves around the Sun precisely in what astronomers call the **"habitable zone"**, where the surface temperature of a planet can maintain liquid water.

There are currently almost 300 planets recognized outside our solar system. Only ten percent of these extrasolar systems seem to have more than one planet. But recently the development and use of a new searching technique called gravitational micro-lensing has made it likely that more planets will be discovered.

> *Is someone out there?*

Gliese 581c and Gliese 581d are the first planets to be discovered outside our solar system that might have liquid water and therefore life. At 191 trillion kilometres (20.4 light-years) away, they are still quite a distance from Earth. We do not have spacecraft that are capable of making this kind of journey. But if we used the fastest space rocket ever built (the Saturn V, used for the *Apollo* missions), it still would take about 600,000 years to get to Gliese 581c or 581d.

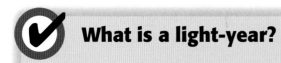

✔ What is a light-year?

Most distances in space are so huge that we measure them in light-years. A light-year is the distance that light can travel in one Earth year. In one light-year, light will travel a distance of 9.46 trillion kilometres (5.88 trillion miles).

Position is everything

It is not just our distance from the Sun that matters for the possibility of life. Also important is what is called the "**galactic habitable zone**", or distance from the centre of a galaxy. If a solar system is too far from the centre of a galaxy, it is likely that its planets lack heavy elements, such as iron, that make life possible. On the other hand, if a solar system is too close to the centre of a galaxy, it risks destruction by a **supernova**, a black hole, or massive amounts of radiation. These habitable zones are like the experience of Goldilocks with the three bears. For Goldilocks, some porridge was too hot, some was too cold, and the last was just right. In the universe, some places are too hot and others are too cold. The position of Earth is "just right". Is it possible that there are other "just right" planets out there? Nobody is sure of the answer to this question, but most astronomers think there could be some other habitable planets in our galaxy, and even more in the entire universe.

> *Is there life on Mars?*

For thousands of years, humans have suspected that there might be life on Mars. Recent investigations, including the mission of the *Spirit of Opportunity* lander shown here, have produced new evidence that Mars probably had surface water at least once in its early history, and probably at other times. This does not necessarily prove that there is, or ever has been, life on Mars, but it is certainly possible. It might even continue to support life far beneath the surface. Some scientists think that life could have originated on Mars and travelled to Earth hundreds of millions of years ago on rocks knocked off by an asteroid impact. So, could we all actually be Martians?

Do other planets actually support life?

Even if there are billions of planets that might potentially support life, there is no guarantee that there is actually life on these far-away worlds. The complex chain of events that led to the formation of life on Earth is still difficult for scientists to understand, let alone replicate. This means there is some uncertainty about whether there is actually life on other worlds. Even if another planet had abundant water and an oxygen-rich atmosphere, it would not necessarily have any life at all – not even a single **bacterium** or plant.

What is astrobiology?

The science of **astrobiology** is the study of life in the universe. It investigates the origin, distribution, and evolution of life with a specific focus on the possibility of extraterrestrial life. Astrobiologists deal exclusively in probabilities, as we have no evidence of any life beyond our planet.

> *Could there be life in our own solar system?*

The recent discovery of very small amounts of atmospheric oxygen on Saturn's frozen moon Titan have created the hope that life might be found there. The frozen surface of Jupiter's moon Europa might be concealing a liquid ocean below in which alien life might find a home.

$$-\frac{\hbar}{i}\frac{\partial}{\partial t} = \frac{p^2}{2m} - \frac{Ze^2}{r}$$

$$\alpha = \frac{\hbar^2}{ec}$$

> Enrico Fermi (1901–1954)

In the early 1950s, physicist Enrico Fermi came up with what has become known as **Fermi's Paradox**. He said that since the universe is so large and so old, if alien life forms existed we should have some evidence of their existence. As we do not have this evidence, he argued, we should conclude that extraterrestrial life does not exist.

Is there intelligent life on other planets?

There may be many millions of planets in our galaxy that support life. Those life forms may even be thriving. But are they *intelligent* life? This is not an easy question to answer. In the first place, we must define what we mean by "intelligent life", a concept that will be covered more thoroughly later. Intelligence is certainly not a key part of life. Consider that on Earth the dinosaurs thrived for many millions of years with tiny brains and no tools or technology. And nobody would describe Earth's most abundant form of life, bacteria, as intelligent.

Some scientists have argued that the development of intelligence is an accidental by-product of **evolution**, while others see it as an inevitable result of evolution. Whichever perspective is true, the fact is that we have no evidence on which to base an estimate of the likelihood of intelligent life. This part of the Drake Equation remains a mystery.

Is anyone trying to communicate with us?

To know if there are other civilizations out there in our galaxy, we would have to have some way to receive a signal from their worlds. This factor in the Drake Equation asks us to estimate the likelihood that alien civilizations would have developed the technology to communicate with other worlds. This technology could include radio or television. If other worlds are listening for evidence of life on Earth, old radio and TV broadcasts could be the first thing they hear from us – even as we speak, the first human signals are travelling through space for distant ears to pick up.

It might seem obvious that other civilizations would have the ability to signal us, but it is not guaranteed. Humans have existed on Earth for about 200,000 years. But we have only used communications that could be received in outer space for just over 100 years, since Guglielmo Marconi (1874–1937) received the first transatlantic radio signal. So, for the vast majority of human history, we have not had the ability to signal other worlds. If the human experience is typical (and we have no idea whether it is or isn't), there could be billions of civilizations in our galaxy, but we would never know it.

> Dr. Frank Drake (born 1930)

Frank Drake was the first person to search for radio signals that might be coming from life forms beyond our solar system. He went on to become one of the founders of the SETI (Search for Extraterrestrial Intelligence) Institute, the most active group now looking for signs of extraterrestrial life.

How long do civilizations last?

The final factor in the Drake Equation is the most ominous. We must consider whether civilizations have a lifespan. If they rise and fall, eventually becoming extinct from technological, environmental, or other factors, then our chances of making contact are reduced considerably. Any number of calamities could mean the end of human society. These range from the self-inflicted to the accidental – such as a catastrophic asteroid collision similar to the one many scientists believe triggered the extinction of the dinosaurs. If there are alien civilizations, they could be the victims of similar catastrophes.

If civilizations have an average and limited lifespan, they may rise up and die out before we have a chance to contact them. Great civilizations could have risen and fallen before humans even evolved on Earth. We simply do not know. But judging from the human experience, societies can do a lot to put themselves in danger – for example, with reckless use of technologies such as nuclear and biological weapons.

> *What is the lifespan of a civilization?*

Even if alien life exists, it may not survive long enough to make contact with us, or human civilization may not survive long enough to make contact with other species. Some people, such as Cambridge University scientist Martin Rees, have argued that humans may not even last another hundred years, due to factors such as nuclear warfare and testing (as shown above), and global warming. Even if the lifespan of a civilization is a million years, that still might not coincide with other civilizations developing and then declining nearby in the universe.

> *Looking for life*

Radio telescope arrays, like this one, are essential in the search for extraterrestrial life. They can be focused on different parts of the sky, scanning different frequencies for signs of life. As they listen to the sounds of the universe, scientists try to discover unusual signals that might be a call from an alien civilization.

How Will We Know?

We're looking … and looking … and looking. But so far, we've found nothing. Then again, we've only looked at a tiny fraction of the galaxy, never mind the larger universe. The search for extraterrestrial intelligence is much more complicated than looking for a needle in a haystack. At least when you're looking for a needle, you know what you're searching for. Also, haystacks are generally of manageable and predictable sizes. There's no such luck with the galaxy, much less the universe. Space is so large that we cannot listen to the whole sky at once. When you add in the fact that we are not even sure exactly what we're looking for, it becomes hard to tell how we will even know when we've received a signal from an extraterrestrial civilization. Something that looks like random blips could be the great work of another society. Our television programmes, streaming beyond the solar system, could be interpreted as nothing more than noise by aliens listening many billions of kilometres away.

But this has not stopped scientists studying and planning the search for extraterrestrial life. Building telescopes and borrowing time on giant receivers, they are listening carefully for any signs of life. In this chapter you will learn about the techniques used. You will also read about a few occasions when some people think they made contact with alien life.

Where should we look?

We are listening very hard for signs of extraterrestrial life. But it's not easy to know where to look. The universe is full of **radiation**. Radiation is emitted when you turn on a light bulb and is created when a plant turns carbon dioxide into oxygen. If other species learned, as we have, to harness the power of radiation for their civilization, they might have technologies such as radio and television – there is thus a possibility that we could use these signals to identify life on other planets.

> *Voyaging into space*

In 1977, the United States launched the first *Voyager* spacecraft. It is now the furthest human-made object from Earth. Almost 16 billion kilometres (10 billion miles) from the Sun, *Voyager* has taken pictures of planets like Jupiter and Saturn, allowing the most comprehensive exploration of our solar system. Onboard, *Voyager* carries a **golden record** of sounds and images designed to convey information about human society to any intelligent extraterrestrial life

But there are billions of radio frequencies, and the universe is humming with activity. Imagine turning the dial on a radio full of jumbled up noises, trying to find a clear signal. This gives you a small idea of what it is like to look for extraterrestrial signals. Fortunately, there is one part of the radio spectrum that has special significance, while also being relatively quiet. This radio frequency, around 1420 megahertz (MHz), is known as the "waterhole", and that is where Dr. Drake and his colleagues began searching in 1960.

But what is so special about 1420 MHz? Firstly, hydrogen (H) molecules emit radiation at this frequency. Nearby on the "dial", a molecule called hydroxyl (OH) transmits its own signal at 1662 MHz. Together, these molecules combine to form H_2O (water), itself essential to the maintenance of life. Scientists believe that extraterrestrials will recognize the significance of this part of the radio spectrum, especially since hydrogen is so widespread in the universe. These days, searches are not limited to the "waterhole" frequency range. There are SETI projects that scan millions of frequencies at the same time. But the "waterhole" is still thought to have special significance to humans, and possibly to alien life as well.

It is not enough to decide what frequency we should search; we also have to identify a part of the sky to focus on for careful listening. Scientists have used different strategies, including looking only at the nearest 100 stars and then only at G-class stars such as our Sun and some K-class stars. But so far, we have not heard from our interstellar neighbours, if we have any.

> *What if alien life doesn't want to be found?*

Very advanced civilizations might be able to conceal their existence from less sophisticated civilizations like ours, and we would never know they existed.

What are we looking for?

The radio spectrum is full of all kinds of noise from all kinds of radiation sources. But some are more likely to be the product of intelligent life than others, because they use combinations of signals that are very unlikely to exist in nature. Radio frequencies are generally loud and full of all kinds of sounds. Sounds that are random or from a known natural source, such as solar radiation, are called noise. Noise is different from a signal. When scientists look at radio frequencies for signs of extraterrestrial life, they scan for some basic kinds of phenomena that can be distinguished from noise in the frequency they are listening to. Some of the things SETI watchers are looking for include:

- spikes in the density of noises in the spectrum
- rises and falls in the transmission power that might mean the telescope is passing over a radio source
- triplets, or three power spikes in a row
- pulsing signals that might be part of a **digital transmission**.

Even if we saw these in a radio broadcast, it would not necessarily mean we had made contact with alien life. In 1965, scientists from the **Soviet Union** believed they had received a signal from aliens, but it turned out that what they had found was a **quasar**, or quasi-interstellar object, a poorly understood mass of matter and energy that emits a lot of radiation:

"Quasars are still only partially understood. Scientists know that they are tremendously powerful sources of electromagnetic radiation and that they are moving away from us at high speeds. They are believed to be extremely turbulent galaxies — a seething mass of matter and energy very different from our own stable Milky Way. It is suspected that at the heart of each quasar lies a black hole, which traps within its intense gravitational field anything that approaches it. As matter and energy are sucked in, but before they disappear behind what physicists call the 'event horizon' (from which there is no return), they collide with other forms of matter already trapped there and emit energy that may just escape the gravitational clutches of the nearby black hole."

[From Michael White, *Life Out There* (New York: Time Warner, 1999)]

It is not clear whether scanning for peculiar radio signals is even the best way to find evidence of alien life. Maybe an advanced alien civilization would see radio as an obsolete technology, like the telegraph. Perhaps they are using technologies we have yet to even imagine. Physicist Freeman Dyson has speculated that very advanced civilizations would directly harness the power of their star by building a giant sphere around it. He has suggested that we could look for the heat generated by such spheres to find advanced civilizations.

 ## SETI and you

Did you know that you can use your home computer to help with the search for extraterrestrial life? The SETI@Home project allows people all over the world to use their Internet-connected computers to download and analyze data from the giant Arecibo telescope in Puerto Rico.

SETI@Home provides a free program that runs in the background on your computer to process the extraordinarily large amount of data generated by radio telescopes scanning the stars for signs of life. When millions of computers are working on the project, they generate much more power than a whole group of supercomputers. SETI@Home is an example of distributed computing, where the computing load for a project is distributed among many different computers. So far, SETI@Home has not produced any evidence of extraterrestrial life. But the project has been a success in showing how distributed computing can work. The network is also used for a variety of other projects, including malaria prevention and computing models of climate change.

Founded in 1984, the SETI Institute is a private organization dedicated to research, education, and public outreach. It is the most important organization in the world in the search for extraterrestrial intelligence. More than 150 scientists work at the SETI Institute.

Are they already here?

While the scientists at the SETI Institute work to scan radio frequencies for signs of life, many millions of people throughout the world believe that we already have plenty of evidence for extraterrestrial life. A recent poll of Americans showed that one third believe that aliens have visited Earth. Sightings of so-called "Unidentified Flying Objects" (UFOs) happen on a regular basis, all over the world. Many of these sightings are attributed to alien spacecraft or other kinds of alien contact.

> **What's that in the sky?**
>
> Many people claim they have already had evidence of alien life through sightings of UFOs like this one. There is no general scientifically accepted evidence of alien visits to Earth, but people continue to be fascinated by the possibility that aliens are already here or are trying to contact us.

In 1966, a sugar cane farmer in Queensland, Australia claimed he saw a saucer-shaped craft rise from his fields and fly away. When he went to investigate, he found a gigantic circle had been made in the field, and the canes were all woven together. To this day, "crop circle" stories like this one are circulated as evidence of alien landings, even though the majority of circles have been shown to be fakes made by humans using rope and boards.

One of the most famous UFO sightings came in 1947 at Roswell, New Mexico. Some people believe that the U.S. Air Force recovered a crashed alien spacecraft there, despite the Air Force's claim that the crash was the remains of a top-secret government experiment with military surveillance equipment.

Although hundreds of people have reported being abducted and even experimented on by aliens, the broad majority of the scientific community has not recognized their accounts as reliable or factual. It is not clear why beings capable of interstellar travel would need to experiment on humans. Other people believe that aliens have visited Earth many times, and they use this to explain the construction of massive and seemingly impossible structures, such as pyramids, all over the world. These accounts are also used to explain the

Nazca lines, a series of gigantic drawings on a high rocky plain in South America that only form identifiable shapes when seen from the sky. While there is no evidence that alien beings were involved in any way in the construction of any of these structures, popular culture continues to be fascinated with the idea of visits from "ancient astronauts". Television series and video games are inspired by the idea that aliens helped to start human society.

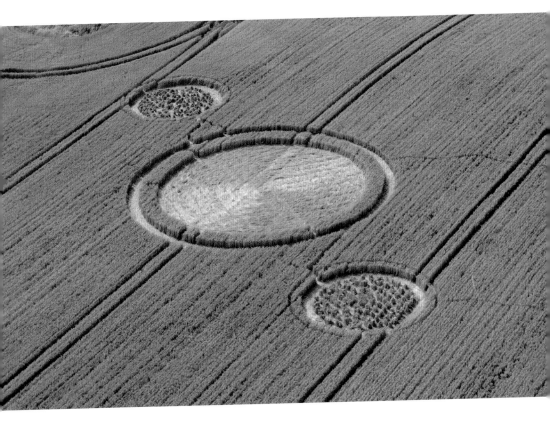

> *Evidence or hoax?*
Giant patterns drawn in fields, like the one shown here, are called crop circles. Many people think they are evidence of alien visits, but most of them are probably hoaxes.

What do you think?

We are fascinated with the idea of contacting an alien race. Do you think other civilizations might use radio frequencies to contact us? Why or why not? Will we be able to detect their signals, even if they do send them? If it is so hard to receive signs of alien life, should we even be trying to find them? Do you think we will ever find evidence of alien life? Or do you think we are alone in the universe?

> *Imagining what's out there*

Humans are fascinated by extraterrestrial life. For all of human history, we have imagined what aliens might look like, conjuring up images of everything from fuzzy, cute aliens to giant, scary ones.

What Will They Look Like?

Creepy or crawly? Fuzzy or scaly? Slimy or leathery? Underwater or in the air? Two legs? Four legs? Eight legs? No legs at all? When it comes to knowing what alien life forms might be like, we have no idea. Or, rather, we have many different ideas and no clue which of them, if any, is likely to be correct. There are so many planets out there, each with their own unique environment and path to life. This means that there could be more forms of life in the universe than we can even dream of. Even on Earth there are more forms of life than we can record, with previously unknown species being discovered all the time. If we cannot even know about all the life forms on our planet, how can we even begin to understand the types of life that might be thriving on faraway worlds?

The question of this book is "Is there other life in the universe?" To formulate a complete answer to the question, it is necessary to think about what it means when we say that something is alive. This makes it easier to consider the different kinds of life that could exist on other planets. Scientists have their own ideas. However, since we do not know what life on other planets might look like, this is a perfect opportunity to use your imagination. Based on what you have read so far, what do you think alien life might be like?

What is life?

Since the beginning of time, something has plagued humanity and all living things. It can exist on its own without eating, drinking, or sleeping for very long periods of time. It doesn't breathe, or grow, or move. But it can kill you. It sounds like an imaginary creature in a science film movie, doesn't it? This thing is the most common form of **genetic** material on the planet. It is one of the simplest and oldest types of thing on Earth, and we're not even sure if it is alive or not. This thing is a **virus**, simply composed of a single piece of genetic material in a wrapper of protein. Viruses remain dormant, or inactive, for long periods of time. When a virus comes into contact with a host creature, it becomes active and uses the host to reproduce its genetic material. Is this virus alive? It depends on what we mean by life.

There is no simple definition that tells us what counts as alive. One narrow definition says that life begins when a creature is self-aware. By this definition, however, someone who has lost brain function would not count as alive, even though their heart was still beating. Also, plants would not count as life, because most people do not think plants have a conscious mind. A more expansive definition holds that anything that has the ability to pass on its genetic material is alive. By this definition, viruses are alive. But is this all there is to life? And if we find viruses on Mars, will we have found life, or simply a chemical grouping capable of reproducing itself?

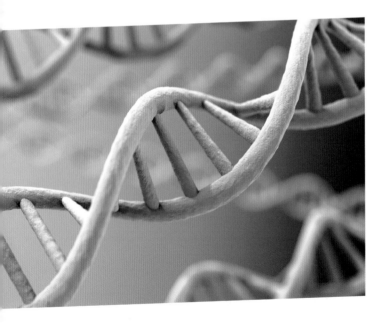

> *The double helix of DNA*

Are the building blocks of life universal? Deoxyribonucleic acid (DNA) is the key to all life on our planet. It holds the instructions for the genes of all known living organisms. But would life on other planets contain DNA? Nobody knows. Most scientists think that alien creatures would have something similar to DNA to transmit genetic information from one generation to another.

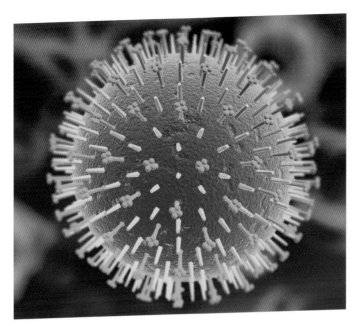

What counts as alive is more complicated than you might think. An entity like a virus can survive without food or water while still reproducing its DNA by using hosts that it infects. But does this count as life? What are the characteristics that define something as alive?

Are Viruses Alive?

"Another way to think about life is as an emergent property of a collection of certain non-living things. Both life and consciousness are examples of emergent complex systems. They each require a critical level of complexity or interaction to achieve their respective states. A neuron by itself, or even in a network of nerves, is not conscious — whole brain complexity is needed. Yet even an intact human brain can be biologically alive but incapable of consciousness, or "brain-dead". Similarly, neither cellular nor viral individual genes or proteins are by themselves alive. The enucleated cell is akin to the state of being brain dead, in that it lacks a full critical complexity. A virus, too, fails to reach a critical complexity. So life itself is an emergent, complex state, but it is made from the same fundamental, physical building blocks that constitute a virus. Approached from this perspective, viruses, though not fully alive, may be thought of as being more than inert matter: they verge on life."

[Luis P. Villarreal *"Are Viruses Alive?"* *Scientific American*, December 2004]

How does life start?

There is no agreement on how life began on Earth. In fact, life on Earth is a relatively recent phenomenon, having only started about 3.85 billion years ago. Until about a billion years ago, the only forms of life on Earth were single-celled organisms. Algae were probably the first multicellular organisms. About 530 million years ago, they were joined by primitive creatures such as jellyfish and primitive worms. Twenty million years later the number of life forms on Earth increased dramatically with what has been called the **Cambrian explosion**. Suddenly, there were many kinds of living thing on Earth, including the invertebrates that gave rise to vertebrates, and eventually to humans.

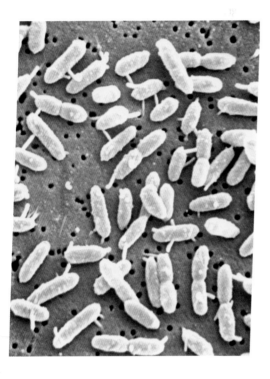

> **> Do aliens look like extremophile bacteria?**
>
> When we think about life on other planets, we might think about the images of aliens we get from films or books – from giant robots to beings that look like variations of humans. But they might not look how we expect at all. What if, when we discover life on other worlds, it is acid-loving **extremophile** bacteria like these, or even viruses?

Nobody knows what caused the Cambrian explosion. One theory is that suddenly evolution found a "way out" of the limited variety of life forms it had been boxed into. Another explanation is that an unusual event, such as a collision between Earth and a giant asteroid, changed the environment so much that new organisms had an opportunity to thrive. Even a small change in the amount of oxygen in Earth's atmosphere could have had dramatic effects on **biodiversity**.

If the creation of life is dependent on unpredictable events, then life might be even rarer than we thought. On the other hand, if evolution proceeds towards diverse and more complex creatures through periods like the Cambrian explosion, alien life might be more numerous than anyone expects.

How does evolution work?

Evolution is a complex process that is essential to explaining the development and maintenance of life. Basically, the theory of evolution predicts that members of a species who are the best suited to their environment will have the best chance of surviving. These creatures will therefore have the best chance of reproducing, spreading their genetic pattern through their offspring. If these offspring inherit genetic material that helps them survive, they will in turn be more likely to pass it on to their offspring.

Genes are constantly **mutating**. Some mutations are minor and cannot even be seen. Others are major – a fish may grow an extra set of fins, or a slug may develop something that is like a backbone. If these mutations give an advantage for survival, the creatures that have them will be the most likely to live long enough to reproduce, passing on the mutation. Eventually, mutations may create a new species. Over time, evolution's random directions, combined with survival benefits, produce the creatures that are best adapted to their environment.

> *Life can thrive in surprising places*

Earth is full of places, like this one, where it seems as if nothing at all could survive. And yet even environments such as this highly acidic river can be full of species that are specially adapted to their environmental niche. If life exists on other planets, could it be just as strange? Could it live in an environment we would never expect to be able to sustain life?

Will they be like us?

Imagine how you think aliens might look. Do you see them as tall and grey, with big eyes? How big are their teeth? Are they scary? Cute? Repulsive? Between films, video games, and books, we are bombarded by images of alien beings. But these creatures are, in some ways, even less strange than the unusual life that shares our own planet with us.

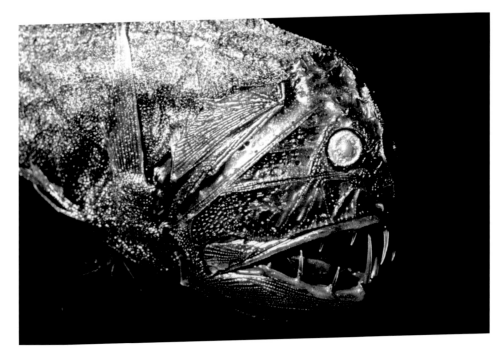

> Alien or Earth creature?

Many people might say this creature could be an alien. Actually, it's a fangtooth, also known as an ogre fish. It was only discovered 50 years ago, in the very deepest parts of the ocean.

Consider the case of *Deinococcus radiodurans*, a **super-resistant** bacterium that can survive acid baths, high and low temperatures, and doses of radiation 3,000 times stronger than would kill a human being, all while hibernating for potentially thousands of years or more. This amazing bacterium is an extremophile, a kind of life form that thrives in extreme environments that might kill other life forms. It is found right here on Earth, but is just about as different from human life as anything we might find in space.

If there is life on other planets, it is not likely to look like us. The chances of life on another planet having the same genetic sequence as humans is about 5 x 10 to the power of 16,557,000, or 5 with more than 16 million zeroes after it.

Little green men?

So what might extraterrestrials look like? If humanity ever travels between stars to explore other planets, it is most likely that we will find bacteria and other simple organisms like the ones here on Earth. But there may be more advanced life forms throughout the galaxy and universe.

Many factors will determine what these life forms look like. For example, alien life might have adapted to breathe carbon dioxide and other gases that humans cannot survive on. This would mean that these creatures would have a very different cellular structure to ours, and would not have DNA (although they might have something similar in function to DNA). If alien life develops on a planet with very different gravity to Earth, for example much lower gravity, there might be massively tall creatures.

> *The "classic grey" alien*

When most people are asked to say what they think aliens look like, they are likely to describe something like the "classic grey" alien. Aliens like these were described by the English novelist H.G. Wells. They became a popular image of extraterrestrial life in the 1960s, after a couple named Betty and Barney Hill described their abduction by aliens that fit the "classic grey" portrayal. Their description of their captors was very similar to aliens that had appeared on an episode of the science fiction television series *The Outer Limits* just two weeks before.

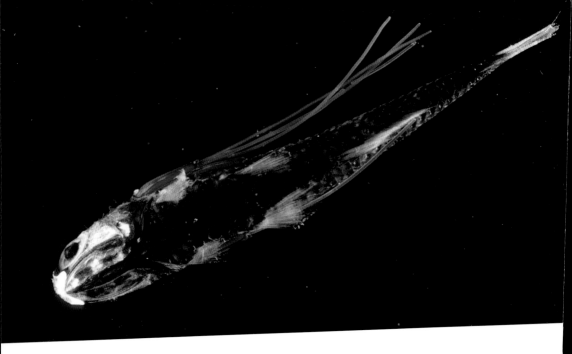

> *What could be more alien than this?*

There is tremendous diversity of life in the oceans and in the far corners of Earth (such as this deepsea lanternfish). Creatures thrive in Arctic cold, inside volcanoes, in rocks buried deep inside Earth, and even in extremely radioactive environments, such as on the control rods of nuclear power plants.

Invisible but intelligent?

Other planets might have life that is not, for whatever reason, likely to try to contact us. With their large brain capacity, dolphins are considered very intelligent. They are also perfectly adapted to take advantage of their environment. But dolphins do not build roads, or spacecraft, or radios, and so are unlikely to be seen by other civilizations looking for signs of life in their galaxy. Could other planets be populated by intelligent life that remains invisible to the galaxy outside? Could it be that humans are unique in their interest in contacting each other and other worlds?

Some scientists have said that the human evolutionary path is likely to be followed by other species in other worlds. They argue that our advances make so much sense that they are likely to be the same advances made by species in other worlds. This line of thinking has been criticized as being "anthropomorphic". This means giving human-like characteristics to things that are not human – as when people attribute human ideas or feelings to dogs or cats. But alien species might be so alien that we cannot even imagine what kinds of values, hopes, dreams, or goals they have.

Three kinds of civilization?

Some SETI researchers have speculated that there might be three kinds of civilization:

- Type I civilizations (which includes human civilization) can use the resources of their own worlds to further their development.

- Type II civilizations can exploit resources beyond their world, and would have mastered interstellar travel. This kind of society would be thousands of years more advanced than ours.

- Type III civilizations would be millions of years ahead of us and could use all the resources available to them in their galaxy.

> *Not what we expect?*

Intelligence may take many different shapes in the universe, just as it takes different shapes among humans. Some humans are peaceful, while some are warlike. If intelligent alien life exists, it might be as strange-looking and bloodthirsty as the aliens from the film *Mars Attack!*, shown here, or gentle and kind. There is simply no way to know.

FROM OUT OF SPACE.... A WARNING AND AN ULTIMATUM!

THE DAY THE EARTH STOOD STILL

MICHAEL RENNIE · PATRICIA NEAL · HUGH MARLOWE

SAM JAFFE · BILLY GRAY · FRANCES BAVIER · LOCK MARTIN

JULIAN BLAUSTEIN · ROBERT WISE · EDMUND H. NORTH

> *Just like in films?*

Humans have long been fascinated by extraterrestrial life. In films like this one, aliens represent a force beyond our understanding that brings a new perspective to Earth. In some films, aliens are our enemies. In others, they are our friends or protectors. But we have no idea what contact with alien life forms might be like. Perhaps this is why we try to represent contact in our popular culture.

What Would We Say To Each Other?

What if we are not alone in the universe? How would we respond to proof that such news was true? How would other life forms treat us? We have no examples of previous extraterrestrial contact to use as evidence, but here on Earth there are plenty of examples of "first contact" to give us some ideas about what human-alien encounters might be like.

When Christopher Columbus landed in 1492 in the area he named the West Indies, the people he found were generally in awe of the large boats and sophisticated technology of their European visitors. Columbus, in turn, found the natives to be extremely generous and kind. However, this meeting of two alien cultures turned out to be anything but a peaceful exchange. Many native peoples were taken back to Spain as slaves for the king and queen. Others were forced to convert to Christianity or face persecution and even execution. Still others died from imported European diseases, such as smallpox, to which they had no **acquired immunity**.

There is no reason to believe that first contact with an alien civilization would necessarily be like this early exchange between the New World and the Old World. How do you think humans might react if they got a message or had visitors from outer space? How would you react?

Could we understand alien communication?

In 1799, the French discovered a curious stone in the Egyptian port city of Rosetta. On it was carved the same text in three different writing systems, including the curious hieroglyphics used by ancient Egyptians, which was poorly understood by antiquaries and linguists at that time. Many people interested in ancient Egyptian civilization had been working for hundreds of years to try to decipher this ancient form of Egyptian picture writing. The text on what became known as the Rosetta Stone allowed researchers for the first time to crack the code of an entirely unknown foreign script by comparing it with known languages.

But what if there is no Rosetta Stone for the entirely alien languages we might encounter if, in fact, we got a message from civilizations beyond our world? Imagine if you were a scientist working late in your lab when you spotted a signal that was unmistakably from an alien race, far away in another solar system. Would you be excited or worried? Would you know how to begin to work out what the message said? Even if you had some success at deciphering it, would you ever be sure you were right?

> *Phoning home*

Many people think that advanced alien civilizations would be peaceful and friendly, having evolved past the impulse for war and bloodshed. Based on the human experience, a civilization would have to survive for a very long time without killing itself off if it wanted to develop interstellar travel. But this does not necessarily prove that alien life forms would be peaceful, like ET from the film.

Some scientists think that there are universal languages we could use to communicate with aliens – for example, with formulas for chemical compounds and other natural phenomena. But how might these scientific ways of speaking translate into areas such as art or music? Would we even be able to understand the hopes, dreams, and fears of a civilization totally unlike ours?

Distance might complicate matters even more. If we received a signal from a civilization that was 60 light-years away, that communication would already be 60 years out of date. Our response would take another 60 years to get back, and then it would take 60 years to receive a response to that response – everyone on Earth involved in sending the response would be dead before we even heard back from our extraterrestrial contacts.

War of the Worlds: broadcasting fear

On 30 October 1938, Martians landed and wreaked havoc in New Jersey, USA. Or so many Americans thought. They had been taken in by a realistic radio broadcast that claimed to tell the unfolding story of an alien invasion. In fact, the broadcast was a fictional programme narrated by Orson Welles. His script was an adaptation of H.G. Wells' novel *War of the Worlds*. The broadcast is probably the single most famous radio production in history.

In H.G. Wells' original novel, Martians land in Woking, England. But Orson Welles moved the invasion to the United States and presented reports on the invasion as a series of fake news broadcasts. It is not actually known how many people really panicked on hearing the fake broadcast, but his presentation made quite an impression on the American people and is credited with making Welles famous.

> *Aliens are here!*

Noted film-maker and actor Orson Welles first made his mark terrifying U.S. radio listeners on 30 October 1938 with his broadcast of a fictitious alien landing.

How would we react?

There is no way to know how humans would react to news that life had been found beyond our planet. Certainly people would probably react differently if the new life was a tiny microbe on the surface of Mars, rather than a giant alien creature landing at the Houses of Parliament in London. Reactions also are likely to be very different depending on religious and political world views. Some religions believe that Earth is special among God's creations. These believers might be stunned by the appearance of other life forms. Other people might see contact with aliens as a sign of the oneness of life, and use the signal to work for peace. But others might have a darker view of first contact, and try to work to exploit alien ideas and technology for their own ends.

The ethics and etiquette of extraterrestrial contacts

An alien civilization with the technology to reach Earth would certainly be vastly ahead of us in technological capabilities. This is why some scientists conclude that contact has not yet been made. They suggest that it is likely that advanced alien cultures, if they are out there, have decided not to interfere with our development as a species. Perhaps humans, as a young species, are not yet ready to join the galactic society. Or perhaps alien cultures are not yet sure of our intentions towards them.

But there is no evidence for this view. We simply have no idea whether life on other planets even exists, much less whether it is peaceful or warlike, advanced or primitive. What we would say to each other and how we would act depends so much on thousands of factors that we do not know or understand, and cannot even speculate about.

> **First contact**

When humans sent the unmanned *Pioneer* spacecraft out beyond our solar system, they hoped to create a moving time capsule, as well as a kind of ambassador for humanity. The spacecraft carried images and sounds thought to represent humanity. If you could create such a representation, what would you include to represent all of life on Earth and human achievement?

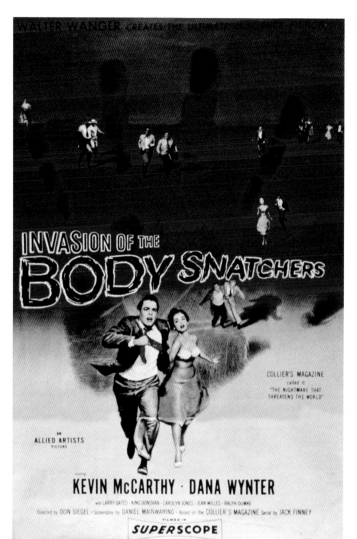

> **Alien world domination?**

Throughout human history, we have been fascinated by the idea of making contact with aliens. During the Cold War, films such as *Invasion of the Body Snatchers* depicted aliens as unfriendly and bent on world domination. This type of alien force was also a metaphor for the Soviet Union and Western fears about **Communism**.

"The prospect of extraterrestrial life offers a new kind of lens through which to see and re-evaluate some of our most basic questions. We are stirred to consider broader ethical perspectives and worldviews. We are prompted to explore how extraterrestrial life might affect not only how we see life and ourselves in the universe, but how we might act in the universe as well. Such prospects are well worth preserving."

[Source: NASA Engineer Mark Lupisella, *"The Value of Martian Microbes."* http://www.pbs.org/exploringspace/ esays/microbes.html]

What do you think? If you could ask an alien civilization questions, what would you ask? What do you think they would want to know from us?

> **Join the debate!**

Do aliens exist? What might they look like? Can we contact them? How
about visiting them? What would we say to each other? This book has raised
a variety of questions that the world's most eminent scientists are debating
even now. When you join the discussion, you are participating in one of the
most captivating issues of all time.

What Do You Think?

Without widely accepted evidence of extraterrestrial life, the idea of life beyond Earth is just a hypothesis. It might be true, and there are good arguments to support the likelihood of alien life, but without direct evidence the idea remains purely hypothetical. This does not mean it cannot be proven to be likely; it just means that you have to be clever about arguing for the existence of extraterrestrial life. Just because you haven't seen something, does not mean it doesn't exist.

In the first part of this book you were introduced to something called "inferential reasoning. Frequently in science it is necessary to draw conclusions about things that cannot be directly observed. Before humans had microscopes sensitive enough to observe DNA, scientists had concluded that cells must have something where inherited material was encoded and transmitted. They were able to infer the existence of DNA without actually seeing it. Later, when DNA was observed, this confirmed their hypothesis!

The debate about whether extraterrestrial life exists is not a simple yes or no question. It has to be addressed as a probability. Without being able to produce an alien body or a message from the stars, we can only say that alien life is likely or unlikely to exist, and support our ideas with reasoning and evidence.

Is there life out there?

Because we do not have any direct evidence of alien life, we have to look at the probability that alien life exists. Probability is a term that means the chances, or odds, of something happening. For example, when you toss a coin, you have a 50 percent chance of it coming up one side or the other. We know the probability because we know that there are two sides, and each one has an equal chance. When we try to estimate the probability of life existing beyond our planet, things get a little more complicated. This book has introduced you to the Drake Equation. It began as an attempt to identify and put numbers to the factors that affect the likelihood of finding an alien civilization. The factors in the Drake Equation include numbers for things we can reliably estimate, like the number of stars and the number of planets. They also include numbers we can only guess at, like the lifespan of a civilization and the likelihood of other civilizations trying to contact us.

What do you think? Is it likely that other life exists somewhere in the universe? Why or why not? Is this a question we can even try to answer with the lack of hard, verifiable evidence?

Given the uncertainty of the issue, it is not surprising that some people think that we should not be using our scarce scientific resources to look for alien signals that may or may not be out there. There is no denying that there are plenty of problems here at home, but natural human curiosity means we want to know what's out there, too. What do you think about this? Is it worth the financial investment needed seriously to look for extraterrestrial life? What if the spending takes money from other scientific research here on Earth, for example, into cures for diseases or ways to create cleaner, sustainable energy sources?

The big issues

It is hard to disprove the idea that aliens exist. At best, the probability might be really, really, really small. But since we simply do not know, we cannot declare the issue settled – especially since the universe is such a big place, with so many corners we have never even seen, let alone explored. If we assume that it is likely that extraterrestrial life exists, then there are many issues raised in this book for further exploration. For example, what is likely to be the nature of alien life? There are as many theories as you can imagine about what alien life might look like. Also, we do not know what other societies might be like. Would they be peaceful or warlike? Would they be like us? Why or why not?

As you begin to think about, research, and debate these issues, you are joining the scientific community's consideration of one of humanity's oldest questions: Are we alone in the universe? Thinking like a scientist about this question will involve looking at the evidence, critically considering the reasoning, and listening carefully to opposing viewpoints as you form and defend your own ideas. One of the most exciting things about this topic is that it is anything but settled. That means that what you think about whether there is life beyond our planet has the potential, any day now, to be proven true or false depending on whether we hear from any of our interstellar neighbours.

> The universe is a big place

Our galaxy, the Milky Way, is so big that it would take more than 100,000 years to cross it travelling at the speed of light. And there are billions of galaxies just like ours all over the universe. Could it be that there are alien civilizations, maybe even some like those on Earth, in our galaxy and beyond? What do you think?

Is there other life in the universe? Let's debate!

There are three important parts of a debate in any format. First, all participants must be able to make arguments to defend their side. Remember that arguments should have three parts: an assertion, reasoning, and evidence (A—R—E). Second, participants should make sure that they respond to opposing arguments. It is not enough just to make an argument for your side; you also have to answer what the other side says. This is called refutation. Third, it is important to take notes during any debate or discussion. This will allow you to track arguments as they are made, prepare to respond to the other side, and organize your ideas for upcoming speeches.

 ## Speaking your mind

Many people are intimidated by the idea of delivering a speech in class or in public. While it is true that public speaking can be difficult, it should be no more difficult than reading a book or writing an essay. It takes preparation and practice. If you are preparing to make a presentation on extraterrestrial life, or on any other issue, a few key steps can make the difference between an uninspired speech and a great one.

✔ *Prepare.* Research your topic thoroughly. Have plenty of facts and arguments to support your position. Make sure your evidence is credible and your reasoning is sound.

✔ *Organize.* Good speeches are organized, and the best way to start is by writing an outline. Your speech should have a thesis statement and supporting points. Your supporting points will be especially effective if organized using the A—R—E method.

✔ *Practise.* You do not need to memorize your speech, but it is very important to practise. Like playing a sport or a musical instrument, public speaking is a skill that requires practice. Try to speak from minimal notes, reducing your outline to a few notecards that remind you of your points. This way you won't be reading from a script and boring your audience.

✔ *Relax.* When you are delivering your speech, try to relax and take deep breaths. It is okay to depart from your script a little bit, or even make a joke. This does not mean you should be unprofessional, but audiences do respond well to speakers who seem friendly.

After your speech, it's a good idea to answer questions from the audience, if appropriate. This gives you a chance to clear up any lingering misconceptions or misunderstandings, as well as to share information with the audience.

> *Are aliens already here?*

If you do personal research into UFO sightings, you will find that there is quite a debate between people who claim to have seen aliens and people who do not believe the evidence. Debating the issues raised in this book can include specific discussions of phenomena such as crop circles and alien abductions, as well as the science of space travel.

Debate formats

Two-sided debate

In this format one side makes a case for the topic. The other side argues against that case. The side arguing for the topic is called the *proposition* or *affirmative*; the other side is called the *opposition* or *negative*. Each speaker on a side delivers a speech. The teams alternate speakers. The proposition team, which must prove that the topic is more likely to be true, speaks first and last. The opening proposition speaker makes a case. The first opposition speaker refutes the case. Second speakers continue with their team's points and refute new points from the other side. The final speeches are summaries of the best arguments for a team and the best refutation against the major points of the other side. With six students, you would have the following format and speaker times:

First speaker, proposition – 5 minutes

First speaker, opposition – 5 minutes

Second speaker, proposition – 5 minutes

Second speaker, opposition – 5 minutes

Third speaker, opposition – 3 minutes

Third speaker, proposition – 3 minutes

It is possible to add question and comment time by the opposing side or by a class or audience during, in between, or after speeches.

Discussion

A group of students participates in a panel discussion on an issue. Students speak for themselves and may agree or disagree with the opinions of others on the panel. The discussion is designed to inform an audience. There is an overall time limit, for example 30 minutes, for the entire discussion. You can use a moderator to ask questions and keep the discussion moving. A panel discussion is an opportunity to use conversation in a way that presents and challenges ideas. Audience questions may be added after the discussion.

Open forum

This is an effective format for a class or large group. A single moderator leads an open discussion on a range of topics. Members of the audience may present new ideas, add to the presentations from others, or refute any issue. Like brainstorming, this format quickly gets a variety of ideas into a discussion.

> *Working together*

When you brainstorm with others and come together to share your research and ideas, you will often find that putting several heads together generates the best arguments. Also, working in a pair or a group helps you prepare for debates as you challenge each other's arguments and search for the very best expression of a particular opinion.

Who has won?

It is possible to ask an individual or group to judge a debate, whatever the format. For larger discussions, the audience can be asked to vote for the team or individuals that did the best job.

No matter what format you use, debating this subject is complex and hard evidence is in short supply. So, are we alone or is there other life in the universe? Have your opinions changed? What do you think?

Find Out More

Books

Planets, Stars and Galaxies, David Aquilou (National Geographic Society, 2007)

Life on Mars, David Getz (Henry Holt, 2004)

Aliens and UFOs, Marc Tyler Nobleman (Raintree, 2006)

Life in Space, Helen Orme (Ransom Publishing, 2008)

Extraterrestrial Life, Tamara L. Roleff (Greenhaven Press, 2001)

Earth's Final Frontiers: Outer Space, Anne Rooney (Heinemann Library, 2008)

Stargazers' Guides: Can We Travel to the Stars? Andrew Solway (Heinemann Library, 2006)

Websites

The SETI Institute Learn about the projects and programmes of the world's most active group looking for extraterrestrial life.
http://www.seti.org

SETI@Home Set your computer up to help look for extraterrestrial life.
http://setiathome.berkeley.edu/

Extraterrestrial Life An archive of articles by *Discover* magazine on the many aspects of the search for extraterrestrial life, this is a starting point for those looking to find out more.
http://discovermagazine.com/topics/space/extraterrestrial-life

Drake Equation Online This interactive version of the Drake Equation allows you to plug in your own numbers and compare them to Dr. Frank Drake's estimates of the number of alien civilizations in the galaxy.
http://www.pbs.org/lifebeyondearth/listening/drake.htm

Films

These films give some film-makers' ideas about what contact with extraterrestrial life might be like.

The Thing From Another World (1951)

Forbidden Planet (1956)

Day of the Triffids (1963)

2001: A Space Odyssey (1969)

The Andromeda Strain (1971)

Close Encounters of the Third Kind (1977)

ET: The Extraterrestrial (1983)

Independence Day (1996)

Mars Attack! (1996)

Contact (1997)

Men in Black (1997)

Galaxy Quest (1999)

War of the Worlds (1953, 2005)

Glossary

acquired immunity immunity to diseases that develops with exposure to those diseases or to specific environmental mechanisms that prevent those diseases

astrobiology science that addresses the question of whether or not there is life in space

bacterium (plural bacteria) organism with a single cell

biodiversity variety of life forms within a habitat or ecosystem

black hole area in space where a very powerful gravitational field means that nothing, including light or radiation, can escape

Cambrian explosion sudden appearance, about 520 million years ago, of many new groups of animals during the geological period known as the Cambrian period of the Paleozoic era

Communism economic and political system which promotes common ownership, rather than private ownership, of property

digital transmission radio signal that uses a binary code (zeroes and ones) to transmit information

Drake Equation equation proposed by SETI pioneer Frank Drake to calculate the number of civilizations in the galaxy. The Drake Equation accounts for a variety of variables that determine the likelihood of extraterrestrial life.

evolution process of change over time in the inherited characteristics of a population of organisms

extraterrestrial beyond Earth. The word is used to refer to alien life forms.

extremophile creature that lives in an extreme environment

Fermi's Paradox named after physicist Enrico Fermi. Fermi's Paradox states that if aliens exist we should have evidence of their existence. As we do not have this evidence, we should conclude that extraterrestrial life does not exist.

galactic habitable zone part of a galaxy most favourable for the existence of solar systems which could support life

genetic to do with genes or DNA, the molecules that encode genes. Genes are the physical basis for inherited traits.

geocentric earth-centred

golden record disc containing information in the form of sounds and images created to give other intelligent life forms information about human society on Earth

habitable zone region of space where conditions are favourable for life of the same type as found on Earth

heliocentric Sun-centred

hypothesis explanation that accounts for a set of facts and can be tested by experiment and investigation

inferential type of reasoning that tries to draw conclusions that reach beyond data that is immediately present

metabolism chemical reactions that occur in living organisms and are essential to the maintenance of life

mutation any change in the DNA of a cell. Mutations can be harmful, beneficial, or have no effect.

quasar quasi-interstellar object that is a poorly understood mass of matter and energy that emits a lot of radiation

radiation energy that occurs in the form of particles or waves

radio telescope type of telescope that "listens" to radio frequencies across a designated spectrum

rare Earth hypothesis theory that the development of complex multicellular life on Earth required a very unlikely combination of factors

solvent liquid that dissolves something

Soviet Union another name for the Union of Soviet Socialist Republics (USSR), a Communist country which existed from 1921 to 1991

super-resistant bacteria and viruses can become resistant to products used to try to kill them, like antibiotics. They are super-resistant when they have developed immunity to most or all remedies.

supernova type of explosion created when a star collapses or rapidly expands

virus tiny infectious agent that needs a host cell in order to grow and reproduce

Index